Mathematics for Economics

Part - 2

HEMANTA SAIKIA

ISBN-13: 978-1517039493

ISBN-10: 1517039495

DEDICATION

To my parents and other family members.

CONTENTS

ACKNOWLEDGMENTS

I would also like to give my special thanks to my family whose patient love enabled me to complete this work. I extend due respect and gratitude to all those who helps and co-operations will be valuable and precious for all time to come before me.

1. Relations

So far as our discussion on the set theory is concerned, the group of appearance of the elements is seemed to be a little importance and we don't care the order in which they appeared. But in the discussion of relations and functions the order of elements takes special role as the orders replicate special meanings of a phenomenon. In this section of the book, we will discuss the order and unordered pairs where the orders of appearance of the elements have special importance.

Ordered Pairs and Unordered Pairs:

In lettering a set say A = {x, y}, we don't think about the order in which the elements x and y are emerged as by definition {x, y} = {y, x}. This is acceptable as the elements of set A are not ordered; for which they are called un-ordered Pairs and the set as un-ordered set. However, if we write two different pairs of set (x, y) and (y, x); with the property of (x, y) ≠ (y, x) unless x = y. This can be also applied to sets with more than two elements. Thus the sets in which the elements have appeared in order are called *ordered sets*.

Example 1: To show the height and age we can form ordered pairs (h, a), in which the first element indicates the height in inches and the second element indicates the age in years. So pairs (6, 65) and (65, 6) will mean different and (6, 65) ≠ (65, 6).

Example 2: In a coordinate plane of four quadrants, the ordered pairs represent different meaning such as (2, 4) is different from (4, 2) where order has different connotation. Here the first elements represent the value of x and the second elements represent the value of y.

Figure: Coordinate Plane of Four Quadrants

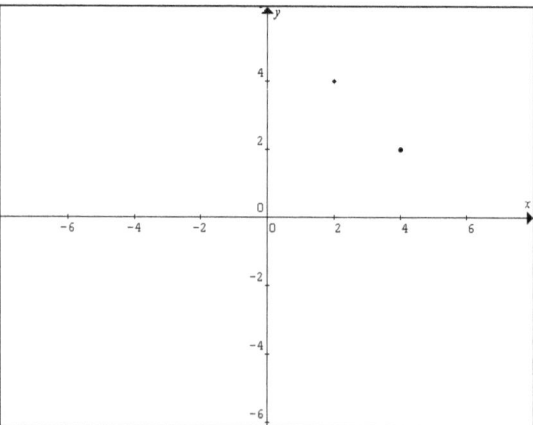

So in coordinate plane, the order of the elements bears special importance and $(x, y) \neq (y, x)$

Cartesian Products

The set of all possible pairs (a, b) where a∈X and b∈Y is called the Cartesian product of x and y and is written as

$$X \times Y = \{(a, b) \mid a \in X, b \in Y\}$$

Thus A ✕ B contains all the ordered pairs of elements in A and B. For three sets:

$$X \times Y \times Z = \{(a, b, c) \mid a \in X, b \in Y, c \in Z\}$$

Example 3: If X = {1, 4} and Y = {x, y}

$$X \times Y = \{(1, x), (1, y), (4, x), (4, y)\}$$
$$Y \times X = \{(x, 1), (x, 4), (y, 1), (y, 4)\}$$
$$X \times X = \{(1, 1), (1, 4), (4, 1), (4, 4)\}$$
$$Y \times Y = \{(x, x), (x, y), (y, y), (y, x)\}$$

Introduction to Relations:

A relation is an association between two or more things, since an order pair indicates the value of y as associates with a given value of x, so the grouping of order pairs will comprise a relation between y and x. The relation provides the value of y, for given value of x. For instance there can be a relation between costs of production (C) and quantity produced (Q) or between revenue (R) of affirm and quantity produced by it. In other language, any subset Cartesian ordered pairs are a relation i.e. any subset of A× B is called a relation between A and B.

Example 4: Suppose, {(x, y)| y = 3x} is set of Ordered Pairs including the values (2, 9), (0, 0) and (-1, -3) which satisfy y = 3x norms. This graph represents a relation between x and y which is depicted bellow.

Figure: Coordinate Plane (y = 3x)

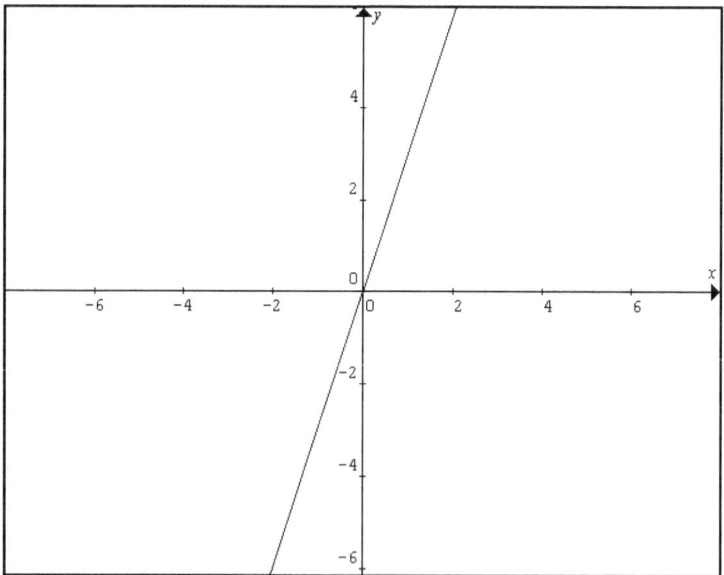

Example 5: Similarly {(x, y)| y ≥ 3x} is set of ordered pairs, depicting a relation between y and x such that the

order pair satisfied the inequality (y ≥ 3x) Such a relation is graphically represented by the set of all points in the shaded area including the points on the line y = 3x as shown in the figure:

Figure: Coordinate Plane (y ≥ 3x)

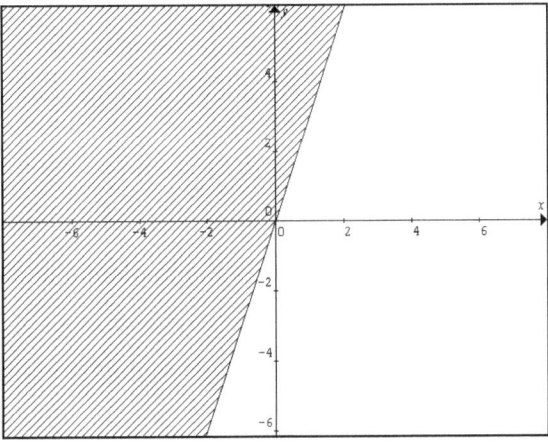

2. Function

A function is a set of ordered a pairs with the property that any value x uniquely determines a y value. The relation in the example: 4 of order pairs $\{(x, y)|\ y = 3x\}$ implies that each value of y is associated with only a unique corresponding value of x. This type of relation between y and x where there for each value of x, there is only one corresponding value of y is called a function. Symbolically:

$$Y = f(x)$$

Generally, 'f' is used to denote the function but some other types of symbols are also used to represents the symbols such as g, h and φ etc. Here, x is called independent variable or explanatory variable and y is called dependent variable or explained variable. On the other hand in the example 4: the relation of order pairs $\{(x, y)|\ y \geq 3x\}$ does not represents a function as there is more than one unique value of y associated with a unique value of x. In economics, the a monopoly firm has no definite supply curve as there is no unique relationship between price and supply of the firm and there may have more than one unique values of price associated with a unique value of supply or output. However, there may be some cases of function where there may be more than one values of x associated with a unique value of y which satisfied the fundamental condition of a function.

Figure: Function

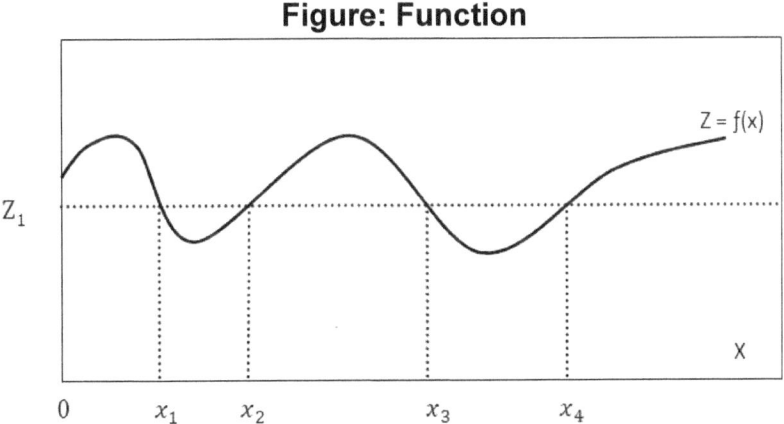

In the above diagram, x_1, x_2, x_3 and x_4 values of variable x are associated with only one unique value z_1 of Z. So it can be termed as function. It is important to note that all the functions represent a kind of relation but all relations may not be formed a function. A function is also called as mapping or transformation as it a work of association of things with one to other.

Domain and Range

If $Y = f(x)$, then the set of all acceptable that x can take in a given context is termed as *domain* of the function. On the other hand set of all y values into which all acceptable x values are mapped is called the *range* of the function. Yet in economics, most of the variables have some kind of restriction such as non negative output, fixed cost etc. for which both the domain and the range has some kind of restriction.

Example: 1: The monthly average cost function [C = $f(Q)$] of a firm is C = 5 + 0.5Q and the firms monthly maximum capacity output is given by 1000 units. Then the domain of the firms cost function will be (0≤ Q ≤ 1000) or

Domain of the function = (0≤ Q ≤ 1000)

And the range of the function will be the 5 (as associated with the lowest value of the Q: 0) to 105 (as associated with the highest value of the Q: 1000).

So, the range of the function = (5 ≤ Q ≤ 100)

Types of Functions

Even though the functions can be divided in to several types depending upon the rule of mapping of functions, but simply we can divide the function into two broad categories:

(1) Algebraic functions;
(2) Non algebraic function.

Algebraic Functions

A function, whose connection with the variable is expressed by an equation that involves only the algebraic operations of addition, subtraction, multiplication, division, raising to a given power, and extracting a given root is called algebraic function. In mathematics, an algebraic function is informally a function which satisfies a *polynomial equation* whose coefficients are themselves polynomials.

▪ Polynomial Function

If $y = f(x)$, then the general form of polynomial function for a single variable say x will be:

$$y = a_0 + a_1 x + a_2 x^2 + a_3 x^3 + \ldots\ldots\ldots + a_n x^n$$

which contains $a_0, a_1, a_2 \ldots\ldots a_n$ are the parameters and non-negative integer powers of variable x. Since the polynomial function has the n highest power of x, so it is termed as *polynomial function* of degree n. We can subdivide the above polynomial functions in to sub classes:

▪ Constant Function

A polynomial function whose range consists of only one specific element or value is called a constant function.

Example: 6 $y = f(x) = c = 10$

Alternatively $y_0 = 10$; this implies whatever is the value of x the y will remain the same i.e. it will remain 10. In economics we will get several cases of constant function such as the demand curve of the firm under perfect completion market will be a constant function. In coordinate plane the curve of a constant function will appears as a horizontal straight line.

Figure: Constant Function

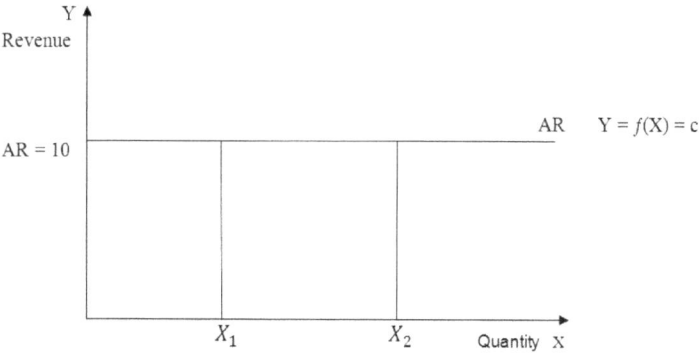

The constant function is a disintegrate case of the polynomial function which contains only the constant part of the function: $Y = a_0$ [a_0 is a constant]

▪ Linear Function

When the power of a variable under polynomial function is equal to one, then the polynomial function is called linear function i.e.

$$y = a_0 + a_1 x \quad [x = x^{(1)}]$$

The linear function will represented by a straight line. However it is not mandatory to have a constant for a linear function but the highest power of the variable must be equal to one. $y = a_1 x$

Figure: Linear Function

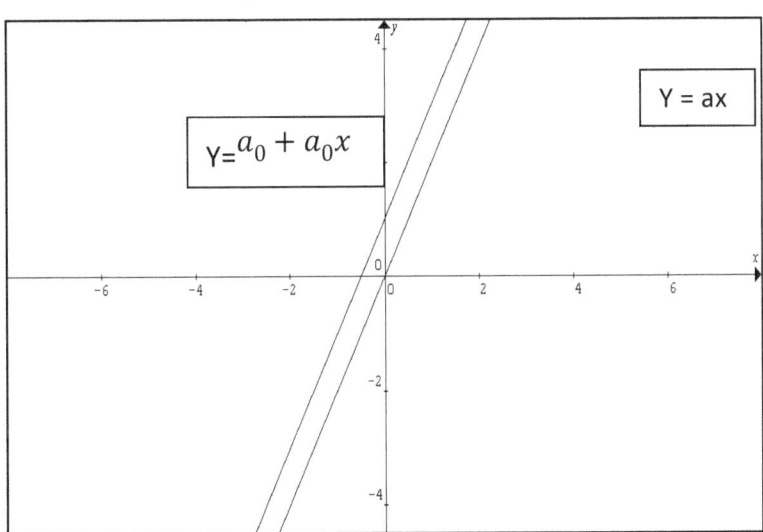

The general for of linear curve will be a straight line with intercept a_0 and slope a_2. The main difference between a linear curve with an intercept and without an intercept is that the former (with intercept) will not pass through the origin and the latter will pass through the origin (with intercept).

▪ **Quadratic Function**

Quadratic function is the second degree polynomial function whose highest power value is 2. If y = $f(x)$, then the general form of quadratic function for a single variable say x will be:

$$Y = a_0 + a_1 x + a_2 x^2$$

Figure: Quadra tic Function

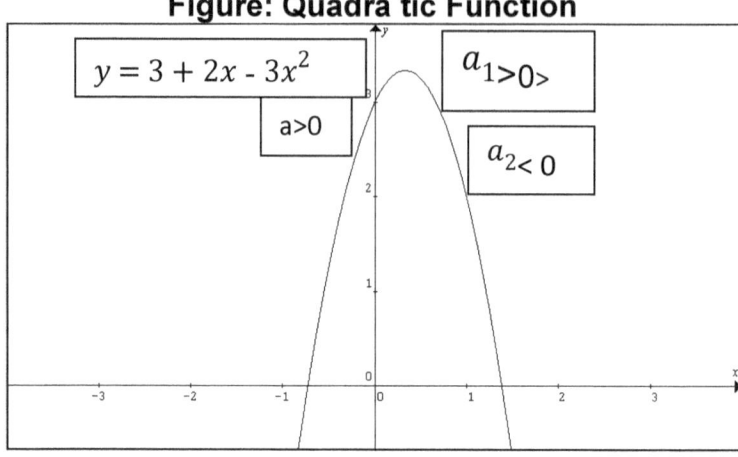

$y = 3 + 2x - 3x^2$

$a > 0$

$a_{1>0>}$

$a_{2<0}$

▪ **Cubic Function**

Cubic function is the third degree polynomial function whose highest power value of variable x is 3. If $y = f(x)$, then the general form of quadratic function for a single variable say x will be:

$$Y = a_0 + a_1x + a_2 x^2 + a_2 x^3$$

The general curve of cubic function shown bellow in the figure:

Figure: Cubic Function

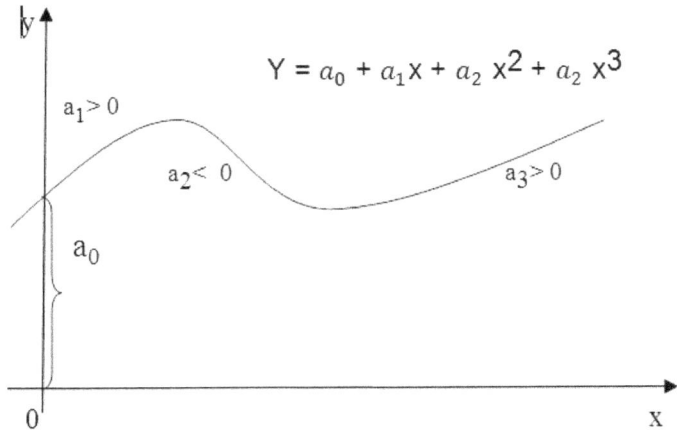

$$Y = a_0 + a_1 x + a_2 x^2 + a_2 x^3$$

$a_1 > 0$

$a_2 < 0$

$a_3 > 0$

a_0

If $Y = 2 + 3x - x^2 - 3x^3$, then the cubic curve will become:

Figure: Cubic Function: ($Y = 2 + 3x - x^2 - 3x^3$)

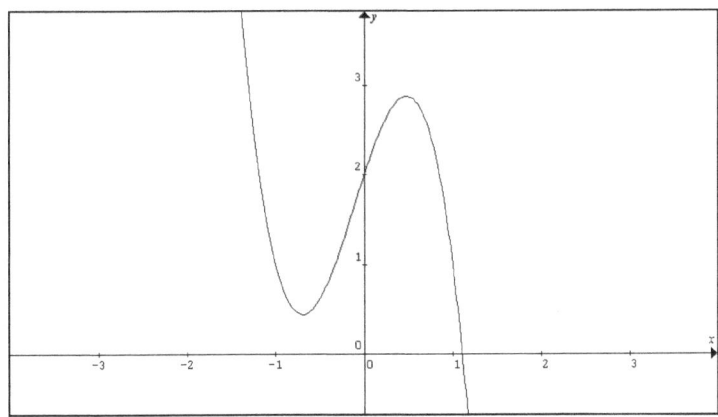

Non Algebraic Function

- ## Rational Function

A function which is expressed as a ration of two polynomial functions is called rational function. The general form of a rational function will be

$$y = \frac{f(x)}{g(x)}$$; Where $f(x)$ and $g(x)$ are two polynomials and $g(x) \neq 0$

Example 1:

$$y = \frac{x - 5}{x^2 + x + 8}$$

Example2:

$$y = \frac{x - 5}{8}$$

It is important to note that according to the definition of polynomial function, every polynomial function can be expressed as a ratio of 1 which is a constant function (Polynomial function). So every polynomial function is a rational function by its definition. There may be some special for of rational function such as

$$y = f(x) = \frac{c}{x}$$ [Where c is a constant]

$$xy = c$$

It is special case of polynomial function which is a ratio of constant function to a linear function with zero intercept. The curve represented by such function is termed as rectangular hyperbola which is convex to the origin. The indifference curve and isoquant are the famous examples of such functions. It is interesting that whatever are the values of x the curve will never touches the x or y axis.

Figure: Rational Function

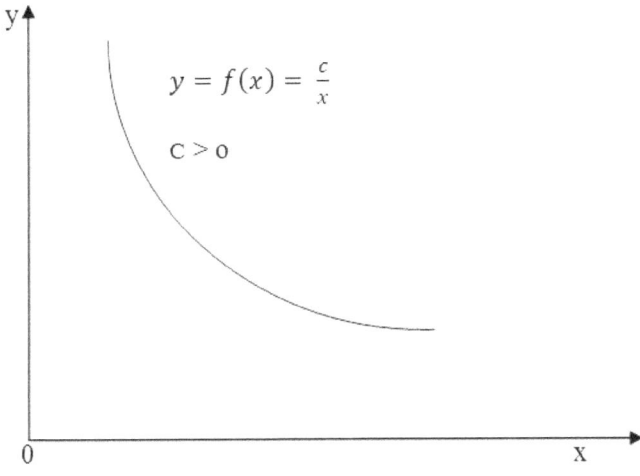

$$y = f(x) = \frac{c}{x}$$

$$c > 0$$

If $y = \dfrac{4}{(5 + 7x)}$, then the curve becomes

Figure: Rational Function: $y = \dfrac{4}{5 + 7x}$

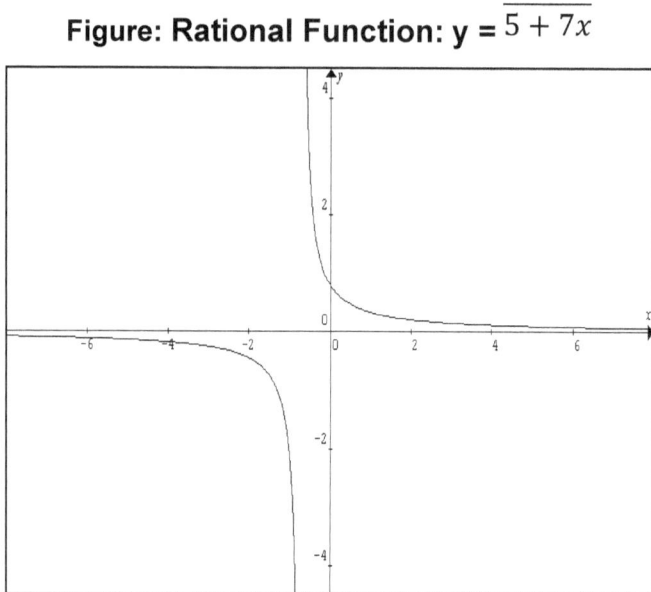

▪ Exponential Function

The function whose independent variable appears as the exponent of a constant is termed as exponential function. For example

$$Y = f(x) = b^x \ (b > 1)$$

The main assumption of b > 1 is taken as if b < 1 and x= 1\2; then y will be imaginary number and if b = 1, then y will be 1 i.e. it will be a constant function form. So b is taken as greater than one.

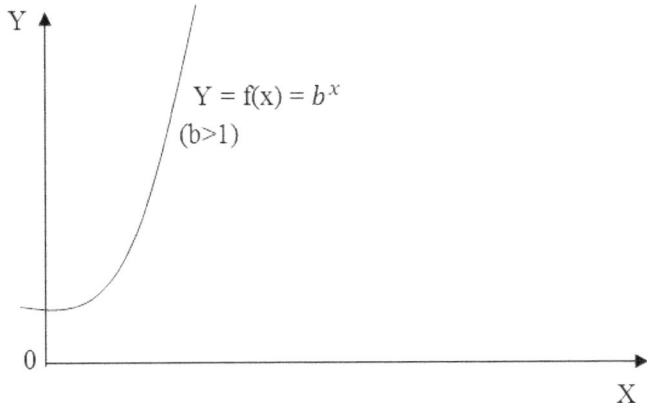

If y = 3^x, then the exponential curve becomes

Figure: Exponential Function: y = 3^x

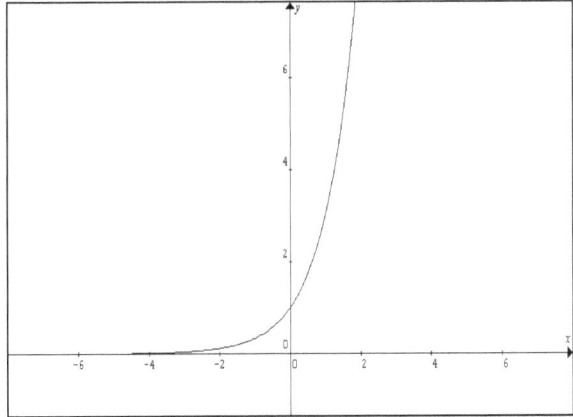

▪ Logarithmic Function

When the dependent variable (y) is a function of the logarithm of the independent variable(x) such a function is termed as logarithmic function such as:

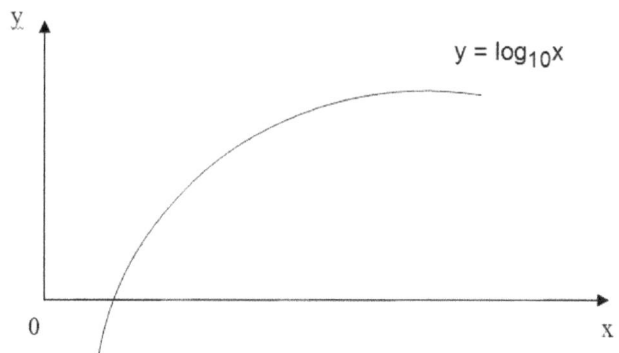

$$y = \log_a x$$

$$y = \log_e x$$

The general shape logarithmic function is shown below:

▪ Some common Rules of logarithmic Function:

- $\log(mn) = \log m + \log n$
- $\log\left(\dfrac{m}{n}\right) = \log m - \log n$
- $\log m^n = n \log m$
- $\log m^m = 1$

Exercise:

1. Identify the order pairs

(i)   A = {x, y, z}   B= {1, 2, 3}

(ii) A = {10, 20, 30}   B= {5, 10, 20}

(ii) A = {10, x, 30}   B= {5, 10, y}

(ii) A = {10, 6, 30}   B= {5, 10, 20}

2. Find the Cartesian products of the following order pairs

(i)   If x = {5, 8}, y = {x, y}

(ii) If x = {a, b}, y = {x, y}

3. What do you mean by relation and function? Differentiate between the relation and function with the help of an example.

4. If c = $\frac{1}{6}$ + 0.80q and the domain is given by (0≤ q ≤ 1000), Find the range of the function.

5. What are the different types of polynomial function?

6. Draw the graph of the following functions:

(a) y = 12   (b) y = $8x$ (c) y = 5+4x (d) $y = 3 + 4x + 3x^2$

(e) $y = \dfrac{7}{2 + 3x}$ (c) $y = y = \dfrac{x^2}{2 + x^2}$ (d) $\pi = -200+50x+0.8x^2 - x^3$

7. Write true or false.

(i) (a, b) ≠ (a, b) unless A = B (where A and B are two sets)

(ii) y = 3x is a function;

(iii) y ≤ 3 is a function;

(iv) y = $a_0 + a_1x + a_2 x^2 + a_3 x^3$ is a polynomial function;

(v) All constant functions are rational function.

(vi) y = 3^x is a logarithmic function.

3. Limit and Continuity of A Function

In mathematics, the limit of a function is a fundamental idea in calculus and in the analysis of concerning the behavior of that function near a particular input. Limits are a mathematical tool which is used to define the limiting value of a function i.e. the value a function seems to approach when it's argument(s) approach a particular value. Although, the argument of the function can be taken to approach any value, limits are helpful in cases where the argument approaches a value where the function is not defined or becomes exceedingly large. In formulas, limit is usually abbreviated as '**lim**'.

Limit of a Function:

Suppose $f(x)$ is a real-valued function (Real-valued function is a function that links to every element of the domain a real number in the image) and k is a real number. The expression will be:

$$\lim_{x \to k} f(x) = l$$

It means that $f(x)$ can be made to be as close to l as desired by making x sufficiently close to k. In that case, we say that "the limit of $f(x)$, as x approaches k, is l". It is noted that this statement is true even if, $f(k) \neq l$. Indeed, the function $f(x)$ need not even be defined at k.

Limit of a Sequence:

The limit of a sequence is one of the oldest concepts in mathematical analysis. It provides a precise definition of the thought of a sequence converging towards the limit. A *sequence* is an infinite ordered list of numbers, for example the sequence of odd positive integers:

1, 3, 5, 7, 9, 11, 13, 15, 17, 19, 21, 23, 25, 27, 29 . . .

Symbolically the *terms* of a sequence are represented with indexed letters: $a_1, a_2, a_3, a_4, a_5, a_6, a_7, \ldots, a_n$

Sometimes we start a sequence with a_0 (index zero) instead of a_1. The sequence $a_1, a_2, a_3, \ldots$ is also denoted by {an} or $\{an\}^{\infty} n=1$.

The limit of a sequence is the value to which its terms approach indefinitely as n becomes large. We write that the limit of a sequence *(an)* is L in the following way:

$$\lim_{x \to \infty} a_n = L$$

For instance, for a sequence

$$1, \frac{1}{2}, \frac{1}{3}, \frac{1}{4}, \dots \text{etc}$$

The limit of a sequence will be:

$$\lim_{n \to \infty} \frac{1}{n} = 0$$

In mathematics, the concept of a "**limit**" is used to describe the behavior of a function as its argument or input either "gets close" to some point, or as the argument becomes arbitrarily large; or the behavior of a sequence's elements as their index increases indefinitely. Limits are used in calculus and other branches of mathematical analysis to define derivatives and continuity.

Left Side Limit and Right Side Limit:

In defining the limit of the function, y =f(x) as: $\lim_{x \to k} f(x) = l$

We say x → k where, variable x can approach a finite number l either from values greater than l or less than l. If x approaches k from the left side i.e. from the values less than k and so, f(x) approaches a finite number l, we termed it as left side limit of x symbolized as $\lim_{x \to k} - f(x) = l$. On the other hand if f(x) approaches l when x approaches k from right side or from values greater than k is termed as right side of limit and is stated as: $\lim_{x \to k} + f(x) = l$.

24

Figure: Left and Right Side Limit

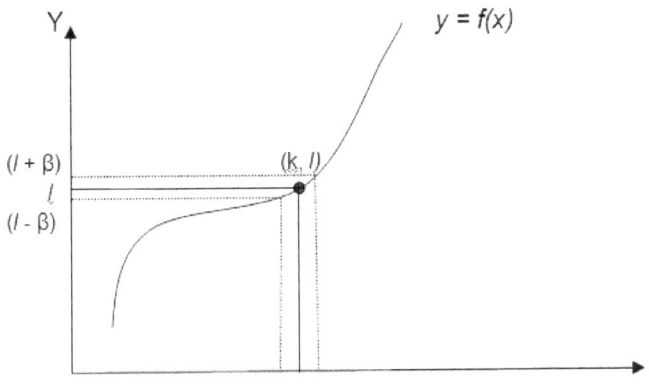

E-3.1: Let a function: $f(x) = \dfrac{x}{x^2 + 1}$ and

$$\lim_{x \to 2} f(x) = 0.4$$

Then the left hand and right hand side limits of the function are

Table: Left and Right Side Limit

$f(1.9)$	$f(1.99)$	$f(1.999)$	$f(2)$	$f(2.001)$	$f(2.01)$	$f(2.1)$
0.4121	0.4012	0.4001	$\Rightarrow 0.4 \Leftarrow$	0.3998	0.3988	0.3882

Left Side Limit   Right Side Limit

So while defining above the function, the $\lim_{x \to 2} f(x) = 0.4$ may approach from both side i.e. from a value of k < 2 or from a value k > 2 which are the two sides of the limit. When the both side limit (Right side limit and left side limit) have a common value say l we state the limit simply as:

$$\lim_{x \to k} f(x) = l$$

(If, $\lim\limits_{x \to k} f(x) = \infty$ we say that the function has no limit or tha function has infinite limit as ∞ is not a number having a definite value so ir is written as f(x)→ ∞; rather than f(x)= ∞)

Evaluation of the Limit of Function:

Generally for getting the limiting value of function we straight forward substitute the limiting value of the function. But straight forward substitution of the value may mislead the function. For example, let a function f(x) = $\dfrac{2 - x}{4 - x^2}$ and we want to evaluate the limit of the function when x→2. Now if we straight forward substitute the value x→2, then the limit of the function will not be defined as the value of the function will be 0. But if we take the limit value after algebraic calculation we get the defined limit value.

(a) $y = f(x) = \lim\limits_{x \to 2} \dfrac{(2 - x)}{4 - x^2}$

$\lim\limits_{x \to 2} f(x) = \lim\limits_{x \to 2} \dfrac{(2 - x)}{(2 + x)\,(2 - x)}$

$$\begin{aligned}&= \lim_{x\to 2} \frac{1}{(2+x)}\\&= \lim_{x\to 2} \frac{1}{(2+2)}\\&= \frac{1}{4}\end{aligned}$$

So after algebraic calculation, the value of the function is found to be 1\4 rather 0

E-3.2. Evaluate the limit of the function:

(b)
$$\begin{aligned}\lim_{x\to 2} f(x) &= \lim_{x\to 2} \frac{2x^2 - 6x + 4}{(2x^2 - 2x - 4)}\\&= \lim_{x\to 2} \frac{2(x^2 - 3x + 2)}{2(x^2 - x - 2)}\\&= \lim_{x\to 2} \frac{(x^2 - 2x - x + 2)}{(x^2 - 2x + x - 2)}\\&= \lim_{x\to 2} \frac{x(x-2) - 1(x-2)}{x(x-2)(x-2)}\\&= \lim_{x\to 2} \frac{(x-2)(x-1)}{(x+1)(x-2)}\\&= \frac{(x-1)}{(x+1)}\\&= \frac{1}{3}\end{aligned}$$

Limit Theorems:

Theorems involving a single Function:

When we have a single function, If y =f(x), the following theorems are applicable:

Theorem 1: If y = αx +β ,

Then, $\lim_{x \to N}(x)$ = αN +β   (α and β are constants)

Example: y = 12x + 23

Then, $\lim_{x \to 3} x$ = 12(3) +23 = 59

Theorem 2: If, y = f(x)=k,

Then$\lim_{x \to N} x$ = k

This implies that the limit of the constant function is constant in that function.

Example: y = 12

Then, $\lim_{x \to 2} y$ = 12

Theorem 3: If, y = x, then $\lim_{x \to N} y$ = N and

If, y = x^m, then $\lim_{x \to N} y$ = N^m

Example: y = x

$\lim_{x \to 2} y$ = 2

Again, $y = x^3$

$$\lim_{x \to 2} y = 2^3 = 8$$

Theorems with Two Functions:

If two functions of the same independent variable x exists such that y = f(x) and w = g(x) and each has a limit

$$\lim_{x \to N} y = L \text{ and } \lim_{x \to N} W = K$$

Theorem 4: (Sum- difference limit theorem)

The limit of the sum (of two functions is equal to the sum of the limits:

$$\lim_{x \to N} [f(x) \pm g(x)] = \lim_{x \to N} f(x) \pm \lim_{x \to N} g(x) = L+K$$

Example: $\lim_{x \to 3} = (x - 3)2$

$= (x^2 - 6x + 9)$

$= \lim_{x \to 3} x^2 - \lim_{x \to 3} 6x - \lim_{x \to 3} 9$

$= 9 - 18 + 9$

$= 0$

Theorem 5: (Product Limit Theorem)

The limit of a constant, c, times a function, f(x), is equal to the constant, c, times the limit of the function:

$$\lim_{x \to N} [f(x) \cdot g(x)] = L.K$$

Example: $(x^2 - x) \cdot (\sqrt{2x})$

$$\lim_{x \to 2} [(x^2 - x) \cdot (\sqrt{2x})] = \lim_{x \to 2} (x^2 - x) \cdot \lim_{x \to 2} (\sqrt{2x})$$

$$= (4-2) \cdot (\sqrt{4})$$

$$= 4$$

Theorem 5: (Quotient Limit Theorem)

The limit of the quotient of two functions is equal to the quotient of their limits, provided the limit of the divisor is not equal to zero:

$$\lim_{x \to N} \frac{f(x)}{g(x)} = \frac{\lim_{x \to N} f(x)}{\lim_{x \to N} g(x)} = \frac{L}{K}$$

[Provided that K≠ 0]

Example: $\lim_{x \to 3} \dfrac{3x^2 + x - 6}{2x - 5}$

$$\lim_{x \to 3} \frac{3x^2 + x - 6}{2x - 5} = \frac{\lim_{x \to 3}(3x^2 + x - 6)}{\lim_{x \to 3}(2x - 5)} = 24$$

Continuity of a Function:

In mathematics, a continuous function is a function for which, intuitively, small changes in the input result in small changes in the output. Otherwise, a function is said to be discontinuous. An intuitive though imprecise (and inexact) idea of continuity is given by the common statement that a continuous function is a function whose graph can be drawn without lifting the pencil from the paper.

When a function y = f(x) posses a limit as x tends to the point M in the domain and when this limit is also equal to y(M) i.e. equal to the value of the function (y = M) and the function is called as continuous function. Thus continuity of a function involves three requirements:

1. The point M must in the domain of the function i.e. *f(M)* is defined.

2. The function must have a limit as x→ M i.e. $\lim_{x \to M} f(x)$ exists;

3. The limit must be equal to f(M) i.e. $\lim_{x \to M} f(x)$ = f(M)

The graph of a continuous function is roughly speaking, a continuous line or a curve with no wavering.

Figure: Continuity of a Function

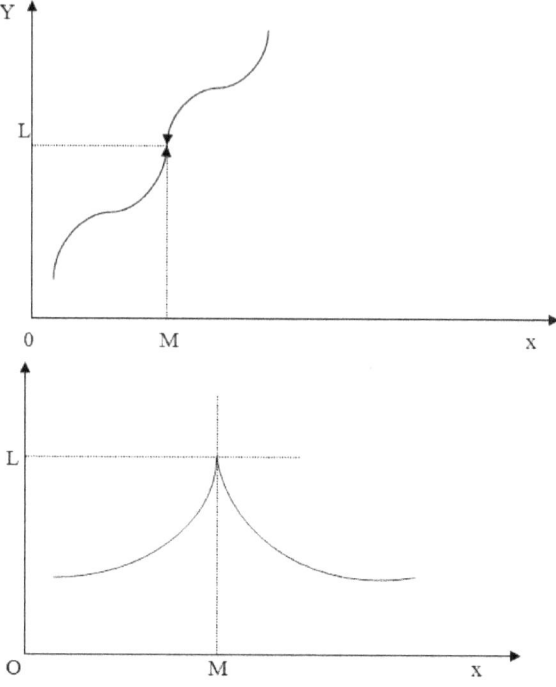

In the diagram all three requirements are met at point M where point M is the domain. and x has the limit L as x→M. At the same time the limit L happens to be the value of the function. So the function is a continuous one. The function in the figure 2 is also a continuous function even though the curve is not a smooth one as when x→M from either side of the curve, y tends to M and $\lim_{x \to M} f(x)$ = L. So the curve is still continuous.

Figure: Discontinuous Curve

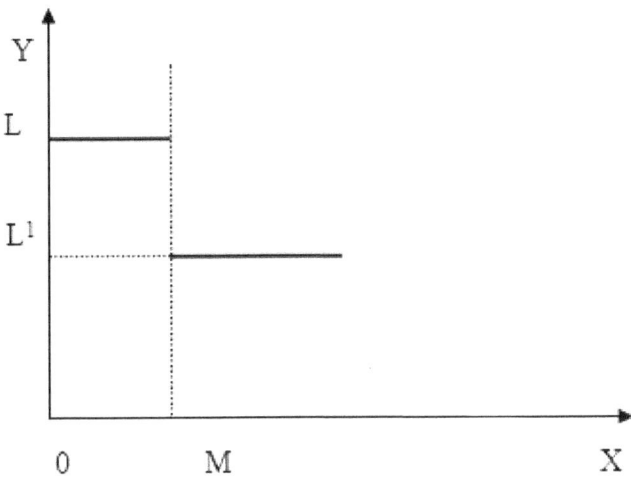

But while we find that a small change in the value of independent variable leads to a sudden skip in the in the value of dependent variable as shown in the figure: 3.1. the curve is not a continues curve at point M as there is no limit exists.

Continuity of a Rational Function:

For a polynomial function y = f(x) we generally found that $\lim_{x \to M} f(x)$ exists and equal to the value of the function at N and since M is a point in the domain of the function. So we can conclude that any polynomial function is continuous in its domain. Since a rational function is the ratio of two polynomial functions. So they are continuous functions in its domain.

List of standard limits:

The following standard limits are helpful in evaluating limits of compound functions. The following listed limits are commonly used for easy reference:

(a) $\lim\limits_{x\to 0} \dfrac{1}{x} = \infty$

(b) $\lim\limits_{x\to 0} \dfrac{\sin(x)}{x} = 1$

(c) $\lim\limits_{x\to 0} \dfrac{\tan(x)}{x} = 1$

(d) $\lim\limits_{x\to 0} \dfrac{1 - \cos(x)}{x} = 0$

(e) $\lim\limits_{x\to a} \dfrac{x^n - a^n}{x - a} = na^{n-1}$

(f) $\lim\limits_{x\to a} \dfrac{\sin(x - a)}{x - a} = 1$

(g) $\lim\limits_{x\to a} \dfrac{\tan(x - a)}{x - a} = 1$

(h) $\lim\limits_{x\to 0} \dfrac{\log e(1 + x)}{x} = 1$

(i) $\lim\limits_{x\to 0} \dfrac{a^x - 1}{x} = \log ea$

(j) $\lim\limits_{x\to 0} \dfrac{e^x - 1}{x} = 1$

Exercise:

1. What is limit of a function? What is left hand side limit and right hand side limit?

2. Evaluate the limit of the following function:

(a) $\lim\limits_{x\to2} \dfrac{(3-x)}{3-x^2}$

(b) $\lim\limits_{x\to3} \dfrac{x^3 - x^2 - 9x}{x^2 - x - 1}$

(c) $\lim\limits_{x\to0} \dfrac{\sqrt{(x+1)} - 1}{x}$

(d) $\lim\limits_{x\to0} \dfrac{\sqrt{(x^2+1)} - \sqrt{(x^2-1)}}{x^2}$

(e) $\lim\limits_{x\to3} \dfrac{x^4 - 2x^3 + 4x - 20}{x^3 + 5x^2 + 3x - 4}$

(f) $\lim\limits_{x\to4} \dfrac{x^2 - 7x + 12}{x - 4}$

(g) $\lim\limits_{x\to1} \dfrac{\sqrt{x^2 + x + 23} - 5}{x - 1}$

(h) $\lim\limits_{x\to5} \dfrac{3x - 15}{\sqrt{(x^2 - 10x + 25)}}$

3. Explain the limit theorems with example.

4. What are the three conditions of the continuity of a function?

5. Write true or false:

(a) $\lim\limits_{x\to0} \dfrac{sin(x)}{x} \neq 1$

(b) If $y = f(x)$ and $\lim\limits_{x\to k} f(x) = l$, we have a continues curve of the function.

(c) $\lim\limits_{x\to a} \dfrac{x^n - a^n}{x - a} = na^{n-1}$

(d) $\lim\limits_{x\to0} \dfrac{1}{x} = 0$